GLOBAL CLIMATE CRISIS

Addressing Urgency and Sustainable Solutions

EDDY MICHAEL

Copyright © 2023 by Eddy Michael

     EDDY MICHAEL

INTRODUCTION

The global climate crisis stands as one of the most pressing challenges of our time, presenting an existential threat to both humanity and the planet's ecosystems. With each passing year, the evidence of climate change becomes more undeniable, manifesting in rising global temperatures, extreme weather events, melting ice caps, and shifting ecosystems. In this introduction, we will provide an overview of the current climate situation, highlighting the urgency for immediate action to mitigate its impacts and secure a sustainable future for generations to come.

Overview of the Current Climate Situation

Over the past century, human activities, particularly the burning of fossil fuels, deforestation, and industrial processes, have significantly increased the concentration of greenhouse gases in the Earth's atmosphere. This buildup of carbon dioxide (CO_2), methane (CH_4), and other greenhouse gases has led to the enhanced greenhouse effect, trapping heat within the Earth's atmosphere and causing temperatures to rise.

The consequences of this warming are far-reaching and profound, affecting every corner of the globe.

One of the most visible signs of climate change is the increase in global temperatures. The Intergovernmental Panel on Climate Change (IPCC) reports that the Earth's average surface temperature has risen by approximately 1.1 degrees Celsius since the late 19th century. This may seem like a modest increase, but it has already resulted in more frequent and intense heatwaves, droughts, and wildfires in many regions.

Furthermore, the warming of the planet is causing the polar ice caps to melt at an alarming rate. Both the Arctic and Antarctic ice sheets are losing mass, contributing to rising sea levels. This phenomenon poses a significant threat to coastal communities, as higher sea levels increase the risk of flooding, erosion, and saltwater intrusion into freshwater sources.

In addition to these changes, climate change is disrupting weather patterns and exacerbating extreme weather events. Hurricanes, typhoons, and cyclones are becoming more intense, while rainfall patterns are becoming more erratic, leading to floods and droughts in different parts of the world.

These extreme weather events not only cause immediate destruction and loss of life but also have long-term consequences for agriculture, water resources, and infrastructure.

Moreover, the impacts of climate change are not limited to the environment; they also have profound implications for human health, livelihoods, and social stability. Vulnerable populations, including the poor, elderly, and marginalized communities, are disproportionately affected by climate-related disasters and disruptions. In many cases, climate change exacerbates existing inequalities and exacerbates poverty, food insecurity, and displacement.

The Urgency for Immediate Action

Given the severity and scale of the climate crisis, there is an urgent need for immediate action to mitigate its impacts and transition to a low-carbon, sustainable future. Delaying action will only exacerbate the challenges we face and make it increasingly difficult to avoid catastrophic outcomes.

First and foremost, reducing greenhouse gas emissions is paramount to limiting global warming and averting the most severe consequences of climate change.

This requires a rapid and significant transformation of our energy systems, transportation networks, industries, and land-use practices. Transitioning away from fossil fuels to renewable energy sources such as solar, wind, and hydroelectric power is essential to decarbonizing our economy and achieving net-zero emissions.

In addition to mitigating emissions, we must also prioritize adaptation efforts to build resilience to the impacts of climate change that are already underway. This includes investing in infrastructure upgrades, disaster preparedness measures, and ecosystem restoration projects to protect communities and ecosystems from extreme weather events, sea-level rise, and other climate-related threats.

Furthermore, addressing the climate crisis requires collective action at all levels of society, from individuals and communities to governments and businesses. Everyone has a role to play in reducing their carbon footprint, conserving resources, and advocating for policies and practices that promote sustainability and environmental stewardship.

Ultimately, confronting the climate crisis demands bold leadership, political will, and international cooperation. Governments, businesses, and civil society must work together to implement ambitious climate policies, mobilize financial resources, and support vulnerable communities in their transition to a sustainable future.

In conclusion, the current climate situation is dire, but it is not insurmountable. By recognizing the urgency of the crisis and taking immediate and decisive action, we can mitigate its impacts, protect the planet, and create a more resilient and sustainable future for all. The time for action is   now.

CLIMATE CHANGE SCIENCE

Understanding climate change requires delving into the intricate mechanisms driving the Earth's shifting climate. Scientific research reveals a clear consensus: human activities, notably the burning of fossil fuels and deforestation, significantly contribute to the unprecedented rise in greenhouse gas concentrations. Through meticulous analysis of temperature trends, atmospheric conditions, and oceanic patterns, climate scientists provide crucial insights into the dynamics of global warming. This knowledge forms the foundation for developing strategies to mitigate climate change and adapt to its inevitable impacts, underscoring the vital role of scientific inquiry in shaping sustainable solutions for our planet.

Understanding Greenhouse Gas Emissions

Greenhouse gas emissions are at the heart of the climate change crisis, representing the primary driver of the warming observed on our planet. To comprehend their role, we must explore the sources, types, and consequences of these emissions.

At its core, the greenhouse effect is a natural phenomenon that keeps the Earth's surface warm enough to sustain life. However, human activities have significantly intensified this effect by releasing large amounts of greenhouse gases into the atmosphere. Carbon dioxide (CO2) is chief among these, primarily originating from the burning of fossil fuels such as coal, oil, and natural gas for energy. Deforestation, industrial processes, and agricultural practices also contribute to the emission of other potent greenhouse gases, including methane (CH4) and nitrous oxide (N2O).

These emissions create a thickening blanket in the Earth's atmosphere, trapping outgoing heat from the sun and preventing it from escaping back into space. As a result, global temperatures rise, leading to climate change. The quantification of greenhouse gas emissions involves complex measurements and modeling, conducted by scientists globally. These efforts help pinpoint the major contributors and formulate strategies to reduce emissions.

Addressing greenhouse gas emissions necessitates a multifaceted approach. Transitioning to renewable energy sources, enhancing energy efficiency, and adopting sustainable agricultural practices are key strategies.

Additionally, carbon capture and storage technologies offer promise in mitigating emissions from industrial processes. A holistic understanding of the sources and dynamics of greenhouse gas emissions is fundamental to formulating effective policies and practices aimed at curbing their detrimental impact on the climate.

Impact on Weather Patterns and Extreme Events

The intricate dance of Earth's weather patterns is undergoing profound changes due to the relentless influence of climate change. As greenhouse gas emissions alter the composition of the atmosphere, the repercussions extend far beyond a simple rise in global temperatures. Understanding these changes requires a closer examination of the shifting dynamics of weather patterns and the increasing frequency of extreme events.

Climate change serves as a catalyst for disruptions in traditional weather patterns, leading to more frequent and severe events. One prominent example is the alteration of precipitation patterns. Some regions experience intensified rainfall, leading to more frequent and severe floods, while others face prolonged droughts, heightening the risk of water scarcity and wildfires. The warming of the atmosphere also contributes to changes in wind patterns, influencing the frequency and intensity of storms and hurricanes.

Rising sea temperatures amplify the destructive potential of tropical storms, leading to more powerful hurricanes and typhoons. The increased heat in the ocean provides the energy necessary for these storms to intensify rapidly, posing a heightened threat to coastal communities. The consequences of these extreme weather events extend beyond immediate damage, affecting agriculture, infrastructure, and human health, with vulnerable populations often bearing the brunt of the impact.

Moreover, climate change has a direct influence on the frequency of heatwaves, which have become more prevalent and intense in various parts of the world. These prolonged periods of excessively high temperatures pose severe health risks, especially for vulnerable populations, and can lead to heat-related illnesses and fatalities. The impact on ecosystems is also profound, with heat stress affecting plant and animal species, leading to shifts in biodiversity and ecosystems.

Understanding the link between climate change and extreme weather events involves comprehensive analysis and modeling by climate scientists. Advanced computer simulations help project future scenarios based on different emission scenarios, enabling policymakers and communities to anticipate and adapt to the changing climate. Initiatives such as early warning systems, resilient infrastructure development, and sustainable land-use planning are crucial components of adapting to the increasing frequency and intensity of extreme weather events.

In conclusion, comprehending the intricate dynamics of greenhouse gas emissions and their impact on weather patterns and extreme events is imperative in addressing the challenges posed by climate change. As we delve deeper into these interconnected processes, the urgency to reduce emissions and implement adaptive strategies becomes even more apparent. The scientific understanding of these phenomena not only informs policy decisions but also empowers societies to proactively mitigate and adapt to the evolving climate landscape.

GLOBAL INITIATIVES

Recognizing the urgency and interconnected nature of the climate crisis, a multitude of global initiatives have emerged to foster collaboration and concerted action on a planetary scale. These initiatives aim to tackle the root causes of climate change, mitigate its impacts, and transition to a sustainable future.

One prominent effort is the Paris Agreement, a landmark international accord adopted in 2015. Signed by nearly 200 countries, the agreement commits nations to limit global warming to well below 2 degrees Celsius above pre-industrial levels, with aspirations to pursue efforts to limit the temperature increase to 1.5 degrees. Countries pledge nationally determined contributions (NDCs) outlining their specific targets and action plans to achieve these collective goals.

The United Nations Framework Convention on Climate Change (UNFCCC) serves as the overarching framework for international cooperation. Annual conferences, known as COP (Conference of the Parties), bring together representatives from nations worldwide to assess progress, negotiate agreements, and strengthen commitments. These gatherings provide a platform for sharing knowledge, technology, and financial resources to accelerate climate action.

In addition to intergovernmental efforts, non-governmental organizations, businesses, and local communities play pivotal roles in driving global initiatives. The Race to Zero campaign, for instance, mobilizes a diverse array of stakeholders, including businesses, cities, regions, and investors, to commit to achieving net-zero emissions by 2050. The Global Green Growth Institute focuses on promoting sustainable, inclusive economic growth through green initiatives and investments.

Financial mechanisms, such as the Green Climate Fund, facilitate the flow of resources from developed to developing nations to support climate mitigation and adaptation projects. This financial assistance is crucial for empowering vulnerable countries and communities to build resilience and transition to low-carbon, climate-resilient pathways.

Collectively, these global initiatives reflect a shared commitment to addressing the complex challenges posed by climate change. While progress is evident, the need for increased ambition, collaboration, and innovation remains paramount. As nations continue to grapple with the evolving climate landscape, these initiatives serve as beacons of hope and catalysts for transformative action on a global scale.

International Agreements and Treaties

The gravity of the global climate crisis necessitates unified efforts on an international scale, leading to the establishment of pivotal agreements and treaties aimed at addressing the root causes of climate change and charting a sustainable future.

One of the most seminal agreements is the Paris Agreement, adopted in 2015 under the United Nations Framework Convention on Climate Change (UNFCCC). Signed by nearly 200 countries, the accord represents a historic commitment to limiting global temperature increases to well below 2 degrees Celsius above pre-industrial levels, with a target to strive for 1.5 degrees. Each participating nation submits nationally determined contributions (NDCs) outlining their specific emissions reduction targets and strategies.

The Kyoto Protocol, an earlier landmark treaty adopted in 1997, laid the groundwork for international cooperation on reducing greenhouse gas emissions. It established binding emission reduction targets for developed countries and introduced market-based mechanisms such as emissions trading and the Clean Development Mechanism (CDM). Although the Kyoto Protocol had limitations, it paved the way for subsequent agreements and provided valuable lessons in the pursuit of global climate goals.

International agreements extend beyond these comprehensive treaties, encompassing a range of initiatives addressing specific facets of the climate crisis. The Montreal Protocol, for instance, focuses on phasing out substances responsible for ozone depletion, indirectly contributing to climate mitigation. The Kigali Amendment to the Montreal Protocol additionally targets hydrofluorocarbons (HFCs), potent greenhouse gases commonly used in refrigeration and air conditioning.

Participation in these agreements underscores a shared recognition of the transboundary nature of climate change. However, challenges persist, ranging from ensuring compliance with commitments to addressing the needs of vulnerable nations disproportionately affected by climate impacts. Persistent diplomatic efforts are required to foster a spirit of cooperation and ensure that nations collectively strive to achieve the ambitious goals set forth in these agreements.

Collaborative Efforts for Climate Mitigation

Collaborative efforts for climate mitigation encompass a broad spectrum of initiatives involving governments, non-governmental organizations, businesses, and communities.

These endeavors aim to reduce greenhouse gas emissions, promote sustainable practices, and foster innovation to transition toward a low-carbon global economy.

Nations worldwide engage in knowledge-sharing and capacity-building through collaborative platforms such as the Intergovernmental Panel on Climate Change (IPCC) and the United Nations Environment Programme (UNEP). These initiatives facilitate the exchange of scientific research, best practices, and technologies, empowering countries to make informed decisions and enhance their climate action strategies.

Multilateral climate finance mechanisms play a crucial role in supporting developing nations' efforts to mitigate and adapt to climate change. The Green Climate Fund, established under the UNFCCC, mobilizes resources from developed countries to assist developing nations in implementing climate projects and transitioning to sustainable development pathways. Such financial collaboration is integral to addressing the global imbalance in responsibility and capacity to combat climate change.

The private sector is a key player in collaborative climate action. Initiatives like the Science-Based Targets initiative (SBTi) encourage businesses to set emissions reduction targets aligned with climate science.

Companies, both large and small, are increasingly recognizing the importance of integrating sustainability into their operations, supply chains, and investments. Partnerships between governments and businesses, such as the Mission Innovation initiative, seek to accelerate the development and deployment of clean energy technologies.

Cities and regions worldwide are forging alliances to share strategies and innovations in urban planning, transportation, and energy efficiency. The C40 Cities network, for example, connects megacities committed to addressing climate change by implementing impactful measures at the local level.

Civil society organizations and grassroots movements play a vital role in holding governments and corporations accountable for their climate commitments. Movements like Fridays for Future, led by youth activists, underscore the urgency of climate action and advocate for systemic changes.

In conclusion, collaborative efforts for climate mitigation are multifaceted, involving diverse stakeholders in a united front against the global climate crisis. As nations and organizations work together to implement sustainable practices, reduce emissions, and innovate for a greener future, the potential for meaningful change becomes not just a possibility but a collective responsibility. By fostering collaboration on all fronts, the world can strive towards a more resilient, equitable, and sustainable future.

          EDDY MICHAEL

SUSTAINABLE SOLUTIONS

Addressing the complex challenges of the global climate crisis necessitates a comprehensive embrace of sustainable solutions that promote environmental stewardship, economic viability, and social equity.

Renewable energy stands as a cornerstone of sustainability. Transitioning from fossil fuels to solar, wind, hydro, and geothermal power is essential to curb greenhouse gas emissions and foster a resilient energy infrastructure. Investments in research and development further enhance the efficiency and affordability of these renewable technologies, accelerating the global shift towards clean energy.

Energy efficiency measures play a pivotal role in sustainable solutions. From industrial processes to residential buildings, optimizing energy use reduces waste, lowers emissions, and contributes to long-term environmental sustainability. Innovations in smart technologies, green building practices, and energy-efficient appliances contribute to a more sustainable and resilient energy landscape.

Sustainable agriculture practices mitigate the environmental impact of food production.

Agroecology, organic farming, and precision agriculture prioritize soil health, reduce chemical inputs, and enhance biodiversity. Embracing circular economy principles in agriculture, such as recycling organic waste and employing regenerative farming techniques, fosters a more sustainable and resilient food system.

Protecting and restoring ecosystems is critical for biodiversity conservation and climate resilience. Afforestation, reforestation, and sustainable land management practices contribute to carbon sequestration and mitigate the impacts of deforestation. Conserving natural habitats and promoting sustainable resource use are integral components of preserving Earth's biodiversity.

Circular economy principles extend beyond agriculture, influencing product design, manufacturing processes, and waste management. Reusing, recycling, and reducing waste not only conserve resources but also diminish the environmental burden of disposal. Circular economy practices create a more sustainable approach to consumption and production.

Engaging local communities in sustainable development initiatives ensures the equitable distribution of benefits and fosters social resilience. Empowering communities through education, capacity-building, and inclusive decision-making processes enhances the effectiveness and sustainability of climate solutions.

 EDDY MICHAEL

In essence, sustainable solutions require a holistic approach that integrates environmental, economic, and social considerations. By embracing renewable energy, enhancing energy efficiency, promoting sustainable agriculture, conserving ecosystems, adopting circular economy practices, and prioritizing social equity, the world can forge a path towards a resilient and sustainable future for current and future generations.

Renewable Energy Technologies

Renewable energy technologies play a pivotal role in the global transition toward a sustainable and low-carbon energy landscape. Embracing these technologies not only mitigates the impact of climate change but also fosters energy security and independence.

1) *Solar Power:* Photovoltaic solar panels harness sunlight to generate electricity. Advances in solar technology, coupled with declining costs, have made solar power increasingly accessible. Large-scale solar farms and distributed rooftop installations contribute to a decentralized and resilient energy grid.

2) *Wind Power:* Wind turbines convert the kinetic energy of wind into electricity.

3) Onshore and offshore wind farms harness the power of the wind, providing a reliable and clean source of energy. Continued innovation in turbine design and offshore technologies enhances the efficiency and scalability of wind power.

4) ***Hydropower:*** Hydropower generates electricity by harnessing the energy of flowing water. From traditional dams to run-of-river and tidal energy systems, hydropower offers a reliable and renewable source of electricity. Balancing environmental considerations is crucial in implementing sustainable hydropower projects.

5) ***Geothermal Energy:*** Geothermal power taps into the Earth's internal heat to generate electricity and heat buildings. Geothermal plants, located in regions with high geothermal activity, provide a constant and baseload source of energy. Enhanced geothermal systems aim to expand the reach of geothermal energy to diverse geological settings.

6) ***Biomass and Bioenergy:*** Biomass, derived from organic materials, can be converted into bioenergy through processes like combustion, gasification, and anaerobic digestion. Bioenergy provides a versatile and renewable energy source, with applications ranging from power generation to biofuels for transportation.

7) ***Emerging Technologies:*** Ongoing research and development focus on emerging renewable technologies such as wave and tidal energy, concentrating solar power, and next-generation biofuels. These technologies hold the potential to further diversify the renewable energy portfolio and address specific energy challenges.

Embracing a mix of these renewable energy technologies is crucial for creating a resilient and sustainable energy system. Integrated policies, incentives, and investments support the widespread adoption of renewables, accelerating the transition away from fossil fuels.

Green Practices in Industries and Agriculture

Green practices in industries and agriculture are integral components of sustainable development, promoting resource efficiency, reducing environmental impact, and fostering resilience in the face of climate change.

Industrial Sustainability

1) ***Energy Efficiency:*** Industries adopt energy-efficient technologies and processes, reducing energy consumption

and lowering greenhouse gas emissions. Combined heat and power systems, advanced manufacturing technologies, and energy management systems contribute to industrial energy efficiency.

2) ***Circular Economy:*** Implementing circular economy principles minimizes waste by recycling, reusing, and recovering materials. Closed-loop manufacturing processes, waste reduction initiatives, and sustainable product design contribute to a more sustainable industrial ecosystem.

3) ***Green Supply Chains:*** Industries are increasingly adopting sustainable and ethical sourcing practices. Green supply chain management involves assessing and improving the environmental and social impact of the entire supply chain, from raw material extraction to product disposal.

Agricultural Sustainability

Regenerative Agriculture: Practices such as cover cropping, crop rotation, and minimal tillage enhance soil health and fertility. Regenerative agriculture focuses on sustainable land management, sequestering carbon in the soil and mitigating the environmental impact of traditional farming.

1) ***Precision   Agriculture:***   Technology-driven   precision agriculture optimizes resource use by employing sensors, GPS technology, and data analytics. This enables farmers to make informed decisions, reducing inputs like water, fertilizers, and pesticides while maximizing crop yield.

2) ***Agroforestry:*** Integrating trees and shrubs into agricultural landscapes enhances biodiversity, conserves water, and improves soil quality. Agroforestry systems offer a sustainable approach to farming that aligns with ecological principles.

Water and Waste Management

1) ***Efficient Water Use:*** Sustainable agriculture and industries prioritize water-efficient practices, utilizing drip irrigation, rainwater harvesting, and water recycling. Efficient industrial water management and conservation efforts contribute to overall water sustainability.

2) ***Waste Reduction and Recycling:*** Industries and farms implement waste reduction strategies, recycling materials, and repurposing by-products. Circular economy principles guide waste management practices, minimizing environmental impact and promoting resource efficiency.

Green practices in industries and agriculture require a holistic approach that balances economic objectives with environmental and social responsibility. Through innovation, technological adoption, and a commitment to sustainability, these practices contribute to building a more resilient and environmentally conscious global economy.

 EDDY MICHAEL

PARTICIPATION OF LOCAL AND COMMUNITY GROUPS IN CLIMATE ACTION

Involving the local and community promotes sustainability, shared responsibility, resilience, and is essential to effective climate action. Involving local communities in grassroots efforts to solve the global climate problem empowers people, increases their ability for adaptation, and supports larger initiatives in this regard.

Education and Awareness in the Community

1) ***Workshops and Outreach Programs:*** To inform the public about climate change, its effects, and the part that each person can play in adaptation and mitigation, local initiatives can host workshops, seminars, and outreach programs.

2) School Programs: Including climate education in the curriculum helps students develop awareness at an early age and prepares them to be environmentally concerned adults.

Initiatives for Local Renewable Energy

1) ***Community solar projects:*** By participating in community solar projects, locals can lessen their dependency on fossil

fuels by pooling their resources to invest in and gain from renewable energy.

2) ***Energy Efficiency Programs:*** To encourage locals to adopt sustainable practices in their homes and businesses, local governments and groups might put in place energy efficiency programs.

Practices for Sustainable Agriculture

1) ***Community Gardens:*** Encouraging the growth of community gardens fosters local food production, sustainable agriculture, and cohesive communities. Additionally, it lessens the carbon footprint brought on by long-distance food transportation.

2) ***Farmers' Markets:*** Participating in and promoting local farmers' markets improves the relationship between farmers and customers, boosts regional agriculture, and lessens the impact of food distribution on the environment.

Urban Planning and Green Infrastructure

Community-led tree planting programs improve green areas, aid in the sequestration of carbon dioxide, and lessen the impact of the urban heat island effect.

- ***Infrastructure for Bikes and Pedestrians:*** Supporting and creating bike lanes, walkways, and public transportation infrastructure lessens dependency on private vehicles and encourages environmentally beneficial forms of transportation.

Initiatives for Recycling and Waste Reduction

1) ***Community Clean-up Events:*** While tackling regional environmental issues like plastic waste, organizing clean-up events promotes a sense of pride in the community.

2) ***Zero-Waste Programs:*** Promoting recycling, composting, and cutting back on single-use plastics are all part of a community's zero-waste strategy.

Building Resilience and Being Ready for Disasters

The creation and execution of emergency plans tailored to a community helps equip its members to withstand the effects of severe weather conditions and other climate-related calamities.

- ***Local Resilience Hubs:*** Creating local resilience hubs helps communities respond effectively to climate-related issues by giving them access to consolidated resources and information during times of crisis.

Litigation and Public Participation

- ***Community Advocacy Groups:*** Establishing or joining neighborhood advocacy groups enables local communities to express their concerns, shape laws, and take part in the deliberations surrounding climate change.

Active engagement with local government structures guarantees the integration of climate considerations into policies, zoning rules, and development plans.

Social and Cultural Initiatives

1) ***Art & Cultural Events:*** Using art and cultural events encourages a sense of social responsibility, sparks innovative solutions, and helps transmit climate concerns in a relatable way.

2) ***Community Celebrations:*** Including climate-related topics in regional festivities and events fosters a good environment, supports sustainable lifestyles, and strengthens a feeling of community commitment.

Through the active engagement of local communities in climate action, we can leverage the combined power of citizens and provide a basis for ecologically conscientious, resilient, and sustainable societies. Local activities play a crucial role in supporting global efforts to mitigate climate change by acting as catalysts for more extensive systemic change.

Environmental Conservation Grassroots Movements

Grassroots movements are strong forces behind environmental conservation because they focus community energy on addressing regional and global ecological issues. These movements, which are frequently started by concerned individuals, are vital in influencing laws, developing sustainable practices, and increasing public awareness of environmental problems.

1) ***Community-Led Conservation Initiatives:*** Local conservation initiatives, like tree planting, habitat restoration, and cleanup campaigns, are started and led by grassroots movements. These initiatives have a direct impact on maintaining biodiversity and enhancing ecosystem health.

2) ***Advocacy for Sustainable Policies:*** Local, regional, and national grassroots groups fight for policies that protect the

environment. These movements interact with legislators in an effort to impact laws and rules that support preservation and sustainable resource management.

3) ***Public Education and Awareness:*** Community education and awareness campaigns on the value of environmental preservation are the main focus of grassroots initiatives. These movements teach people about sustainable living habits and the effects of their choices on the environment through workshops, seminars, and awareness campaigns.

4) ***Community-Based Monitoring and Research:*** Through citizen science programs, grassroots movements frequently give local residents the opportunity to take part in environmental monitoring and research. This targeted data collecting promotes evidence-based lobbying and improves knowledge of regional environmental challenges.

Grassroots movements serve as a first line of defense against initiatives that pose a threat to the environment. These movements use community resilience to safeguard the environment, whether they are opposing hazardous development projects, demonstrating against pollution, or fighting deforestation.

1) ***Networking and Collaboration:*** Local communities, NGOs, and companies are just a few of the many

stakeholders that grassroots organizations encourage to work together. These networks increase the effectiveness of group action by fostering a synergistic approach to addressing environmental concerns.

2) ***Initiatives for Environmental Justice:*** Grassroots initiatives frequently promote environmental justice by addressing inequalities in the allocation of environmental benefits and costs. They seek to prevent disadvantaged groups from suffering unduly from environmental deterioration and to provide them a say in decision-making procedures.

3) ***Campaigns for Behavioral Change:*** By promoting sustainable behaviors on a personal and communal level, grassroots projects stimulate behavioral change. This entails encouraging energy conservation, trash minimization, and conscientious consumption practices.

4) ***Artistic and Cultural Engagement:*** Grassroots movements use music, art, and culture to powerfully communicate environmental themes. Creative expression encourages a sense of responsibility and builds emotional ties to the natural world.

5) ***Global Solidarity and Awareness:*** By collaborating with like-minded projects around the world, grassroots movements help to raise awareness of environmental issues

on a global scale. These movements spread success stories, tactics, and narratives via social media and other channels, strengthening the global environmental movement's sense of unity.

Essentially, grassroots movements represent the idea that protecting the environment is a shared duty. These movements make a substantial contribution to the overarching objective of establishing a sustainable and peaceful coexistence between humans and the environment by uniting communities and encouraging a sense of ownership over regional ecosystems.

Education's Contribution to Increasing Climate Awareness

In order to shape social reactions to climate change as well as attitudes and behaviors, education is essential. Building a sustainable future, encouraging action, and raising climate awareness all depend on an informed and educated public. From formal education institutions to community-based learning initiatives, education plays a multifaceted role in increasing climate awareness.

1) ***Integration into Curriculum:*** Increasing public understanding of climate change is a major responsibility of formal education institutions, ranging from elementary schools to universities. Students are better able to comprehend the science, effects, and solutions associated with climate change when themes linked to it are included in the curriculum. Students that receive environmental education are empowered to become environmentally conscious citizens and are also given a sense of responsibility.

2) ***Outdoor education and experiential learning:*** Students get firsthand experience with nature and environmental issues through field trips and outdoor education programs. Through these encounters, they develop a deeper comprehension of ecosystems and a sense of environmental care.

3) **Teacher Training and Professional Development:** To effectively teach issues related to climate change, educators require support and training. Programs for professional development give educators the skills and materials they need to incorporate climate-related subjects into their teaching, guaranteeing that their students learn current, correct information.

4) **Including Climate Literacy:** Climate literacy is an essential part of education. It comprises knowledge of the scientific underpinnings of climate change, its effects on society, and measures for adaptation and mitigation. People with this literacy are better able to evaluate information critically and make wise decisions.

5) ***Community-Based Education Initiatives:*** In addition to formal education, workshops, adult education courses, and community-based initiatives can increase public knowledge of climate change. These programs provide communities

the tools they need to take on local environmental issues and implement sustainable practices.

6) ***Digital Learning Resources:*** Access to climate-related knowledge is improved by utilizing technology, such as educational apps, online platforms, and virtual simulations. For students of all ages, digital materials offer dynamic and interesting opportunities to investigate climate science and solutions.

7) ***Cross-Disciplinary Approaches:*** By including climate change education into the study of science, social studies, literature, and the arts, among other subjects, students are better able to understand the complex nature of climate challenges. This interdisciplinary approach promotes a comprehensive comprehension of the interdependence of environmental issues.

Encouraging youth activism, environmental clubs, and student-led activities can raise awareness of climate change and produce a generation of climate-aware people who actively support sustainable practices and climate action.

1) ***Collaborations with Nonprofits and NGOs:*** Academic institutions may engage in cooperative efforts with non-

governmental organizations and nonprofits that focus on environmental education. Partnerships ensure a comprehensive approach to climate awareness by expanding the reach and effectiveness of educational activities.

2) ***Continuous Learning for Lifelong Impact:*** Learning about climate change is something that happens all the time, not only in formal education. Programs for adult education, workshops, and open talks promote a culture of lifelong learning and continuing climate awareness.

To sum up, education plays a vital role in increasing awareness of climate change by influencing behavior, forming values, and fostering a sense of responsibility for the environment. Societies may lay the groundwork for informed decision-making and coordinated action in the face of the global climate problem by giving priority to climate education at all levels.

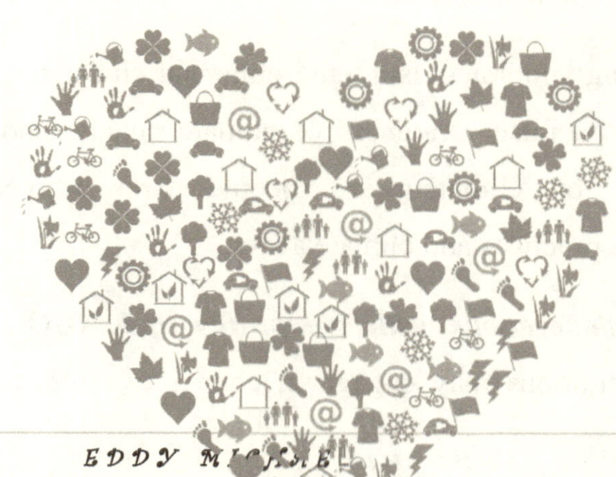

  EDDY MICHAE

GOVERNMENT POLICIES

Effective government policies are crucial in addressing the issues of climate change, fostering sustainability, and transitioning towards a low-carbon future. These policies, adopted at local, national, and international levels, play a crucial role in defining regulations, incentives, and frameworks that influence both public and private sector operations. Here are important areas where government policies can contribute to climate action:

Renewable Energy Targets and Incentives

1) ***Setting aggressive Goals:*** Governments can establish clear and aggressive targets for the adoption of renewable energy sources, aiming for a specified percentage of energy generation from renewables by a certain date.

2) ***Financial Incentives:*** Offering financial incentives, such as tax credits, subsidies, and feed-in tariffs, encourages businesses and individuals to invest in and utilize renewable energy technologies.

Carbon Pricing and Cap-and-Trade Systems

1) ***Carbon Taxes:*** Implementing a carbon tax puts a price on carbon emissions, providing a financial disincentive for enterprises and individuals to create greenhouse gases.

Revenue earned from the tax can be reinvested in sustainable activities.

Cap-and-Trade Systems: Establishing cap-and-trade systems imposes a limit on overall emissions and allows industries to trade emissions allowances, boosting efficiency in emission reductions.

Energy Efficiency Standards:

Building Codes: Governments can enforce tight energy efficiency standards for new constructions, renovations, and appliances, reducing overall energy use in residential and commercial structures.

Transportation Standards: Implementing fuel efficiency standards for automobiles and supporting the usage of electric vehicles contribute to decreasing emissions from the transportation sector.

Protection and Restoration of Ecosystems:

Afforestation and Reforestation: Policies promoting afforestation and reforestation efforts help to carbon sequestration, biodiversity conservation, and the preservation of essential ecosystems.

2) ***Wetland Conservation:*** Protecting and restoring wetlands acts as a natural carbon sink, enhances water quality, and promotes biodiversity.

International Cooperation and Agreements

1) ***Commitment to Global Agreements:*** Governments can actively participate in and uphold international agreements like the Paris Agreement, indicating a commitment to collective climate action on a global scale.

2) ***Technology Transfer and funding:*** Supporting international cooperation through technology transfer and funding systems helps poor nations implement sustainable solutions and adapt to climate consequences.

Adaptation and Resilience Planning

1) ***National Adaptation Plans:*** Developing and implementing national adaptation plans helps governments prepare for and respond to the impacts of climate change, concentrating on increasing resilience in vulnerable people and sectors.

2) ***Infrastructure Resilience:*** Integrating climate resilience concerns into infrastructure development ensures that new and existing infrastructure can endure the changing climate.

Investments in Research and Innovation

1) ***investment for Clean Technologies:*** Government investment for research and development of clean and sustainable technologies fosters creativity and speeds the transition to a green economy.

2) ***Innovation Challenges:*** Governments can establish innovation challenges and competitions to incentivise the private sector, universities, and startups to discover innovative solutions for climate challenges.

Circular Economy Initiatives

1) ***trash Management Policies:*** Governments can enact policies that promote trash reduction, recycling, and the circular economy. This includes initiatives to limit single-use plastics, increase product recycling, and minimize landfill waste.

2) ***Extended Producer accountability:*** Placing accountability on product makers for the whole life cycle of their products fosters sustainable design, materials, and end-of-life management.

Public Transportation and Sustainable Mobility

1) ***Investment in Public Transportation:*** Governments can invest in and promote public transportation infrastructure,

making it more accessible and attractive, hence reducing reliance on individual cars and lowering emissions.

2) ***Cycling and Walking Infrastructure:*** Developing infrastructure for cycling and walking fosters sustainable, low-emission modes of transportation.

Public Awareness and Education Campaigns

1) ***Climate Education Programs:*** Implementing climate education programs in schools and communities promotes public knowledge and comprehension of climate change, building a culture of sustainability.

2) ***Media and Communication Strategies:*** Governments can harness media and communication channels to spread information about climate challenges, policies, and individual actions, encouraging a broader societal awareness and engagement.

To be effective, government strategies for climate action require comprehensive consideration of social, economic, and environmental concerns. Additionally, they should be dynamic, adjusting to growing scientific knowledge and technology breakthroughs to ensure a robust and sustainable future. Public participation, transparency, and collaboration with varied stakeholders are crucial aspects for the successful implementation of these programs.

Legislative Measures for Climate Change Mitigation

Legislative measures are key tools for governments to establish laws that address climate change mitigation. These policies create a legal framework, define standards, and allocate resources to drive the transition towards a low-carbon and sustainable future.

Renewable Portfolio Standards (RPS)

- ***Mandating Renewable Energy:*** Legislating RPS mandates a particular percentage of a region's energy mix to come from renewable sources. This policy incentivizes the development and integration of renewable energy into the grid.

Carbon Pricing

- ***Carbon Taxes and Emissions Trading:*** Legislating carbon pricing mechanisms, such as carbon taxes or cap-and-trade systems, places a price on carbon emissions. This offers economic incentives for firms to minimize their carbon footprint and invest in greener technologies.

Energy Efficiency Standards

- ***Building Codes and Appliance Standards:*** Enforcing high energy efficiency standards for buildings and appliances guarantees that new projects and products fulfill environmentally friendly criteria, reducing overall energy consumption.

Fuel Efficiency Standards

- ***Automobile Emission Standards:*** Legislation defining fuel economy standards for automobiles stimulates the development and acceptance of electric and hybrid vehicles, lowering emissions from the transportation industry.

Investment in Sustainable Infrastructure

- ***Legislative Funding and Incentives:*** Governments can establish legislation to allocate funds and provide tax incentives for the construction of sustainable infrastructure projects, including public transit, renewable energy facilities, and smart cities.

Deforestation Prevention and Afforestation Legislation

- ***Protection of Forests:*** Legislating measures to avoid deforestation and encourage afforestation assures the maintenance of key carbon sinks and biodiversity, helping to climate change mitigation.

Waste Management Legislation

- ***Circular Economy Laws:*** Legislation fostering a circular economy provides requirements for waste reduction, recycling, and sustainable product design. Extended producer responsibility rules make producers accountable for the life cycle of their products.

Research and Innovation Funding

- ***Legislation for Research Grants:*** Governments' can establish legislation to allocate funds for research and development in clean energy technology, sustainable agriculture practices, and innovative solutions for climate change mitigation.

Building Climate Resilience Legislation

- ***National Adaptation Plans:*** Legislation can require the development and implementation of national adaptation plans, ensuring that countries are prepared for and robust to the impacts of climate change.

Community Engagement Legislation

- ***Legal Recognition of Community Initiatives:*** Legislating the recognition and support of community-led initiatives for environmental conservation and sustainable practices supports grassroots movements and local climate action.

International Cooperation and Commitments

- ***Legislative Support for International Agreements:*** Governments can pass legislation that supports the country's obligations to international agreements like the Paris Agreement, ensuring that climate goals are integrated in national policies.

Education and Awareness Legislation

- ***Integration of Climate Education:*** Legislation can mandate the inclusion of climate change education in school curricula, fostering awareness and understanding of climate challenges from an early age.

Green Building Legislation

Requirements for Sustainable Construction: Enacting legislation that enforces green building methods, such as LEED certification, stimulates the construction of energy-efficient and environmentally friendly facilities.

Carbon Capture and Storage (CCS) Legislation

- **Regulations for CCS Implementation:** Governments can establish legislation to regulate and incentivize the deployment of carbon capture and storage systems in industries with high emissions.

Incentives for Sustainable Agriculture

- **Legislation Supporting Regenerative techniques:** Laws can be adopted to incentivise and promote sustainable and regenerative agriculture techniques that absorb carbon, reduce emissions, and boost soil health.

Enacting and enforcing these legislative measures involves collaboration across government ministries, involvement with stakeholders, and continual monitoring and assessment to guarantee efficacy in attaining climate change mitigation goals.

Assessing the Effectiveness of Environmental Policies

Effectively tackling environmental concerns needs regular examination of policies to ensure they achieve their intended objectives. Assessing the success of environmental policies entails a detailed and methodical study of their effects on the environment, society, and the economy. Key considerations for such reviews include:

Environmental Indicators

1) ***Air and Water Quality:*** Monitoring changes in air and water quality gives concrete indicators of the efficacy of programs aimed at decreasing pollution and safeguarding ecosystems.

2) ***Biodiversity Metrics:*** Assessing the state of biodiversity and ecosystems helps quantify the success of conservation programs and habitat protection initiatives.

Economic Impact

1) ***Green Job Creation:*** Evaluating the growth of green industries and the creation of environmentally friendly jobs

provides insights into the economic impact of policies supporting renewable energy and sustainable practices.

2) ***Cost-Benefit Analysis:*** Conducting cost-benefit studies helps analyze whether the economic benefits of environmental policies outweigh the costs, considering both short-term and long-term implications.

Social Equity and Justice

1) ***Distribution of Benefits and costs:*** Assessing how environmental regulations influence different social groups ensures that the benefits and costs are dispersed equally, eliminating disproportionate impacts on vulnerable areas.

2) ***Community Engagement:*** Evaluating the extent of community engagement and participation in policy creation and implementation ensures that local viewpoints are considered.

Regulatory Compliance

1) ***Industry Adherence:*** Monitoring industry compliance with environmental norms and standards is vital for measuring the success of programs aimed at decreasing emissions, pollution, and resource consumption.

2) ***Enforcement Measures:*** Evaluating the efficiency of enforcement measures, such as penalties for non-

compliance, assures the accountability of companies subject to environmental legislation.

Technological Adoption

1) ***Uptake of Clean Technologies:*** Assessing the adoption and integration of clean and sustainable technologies by industries and businesses demonstrates the success of policies fostering technological innovation.

2) ***Investments in Research and Development:*** Tracking investments in research and development in environmentally friendly technologies provides insights into the policy's impact on innovation.

Public Awareness and Behavior Change

1) ***Surveys and Feedback:*** Conducting surveys and getting feedback from the public helps measure the level of knowledge and understanding of environmental issues, as well as the success of communication strategies.

2) ***Behavioral Change Indicators:*** Assessing changes in public behavior, such as adoption of sustainable practices and reduction in resource use, reveals the influence of policies on individual activities.

Long-Term Sustainability

1) ***Resilience to Climate Change***: Evaluating the resilience of ecosystems, infrastructure, and communities to climate change impacts measures the long-term sustainability of policies.

2) ***Consistency with Global Goals:*** Aligning policy outcomes with global sustainability goals, such as the Sustainable Development Goals (SDGs), ensures a comprehensive approach to environmental protection.

Adaptability and Flexibility

1) ***Review and Revision Processes:*** Regularly assessing and rewriting policies based on changing environmental

conditions, scientific developments, and societal requirements promotes adaptation to emerging issues.

2) ***Stakeholder Consultation:*** Engaging with varied stakeholders in the evaluation process ensures that multiple viewpoints are considered and supports continual development.

Transparency and Accountability

1) ***Reporting systems:*** Establishing transparent reporting systems enables for the open assessment of policy outcomes, encouraging accountability and trust in the decision-making process.
2) ***Independent Audits:*** Conducting independent audits and evaluations by third-party organizations gives credibility to the assessment process and ensures unbiased evaluations.

Interconnectedness of Policies

- ***Consideration of Synergies and Conflicts:*** Evaluating how different policies interact with each other ensures that they complement rather than contradict one another, creating a cohesive and effective regulatory system.

International Collaboration

- ***compatibility with Global Efforts:*** Assessing the compatibility of national environmental policies with international agreements, such as the Paris Agreement, ensures coordinated efforts towards global environmental goals.

Assessing the success of environmental policy is an ongoing and dynamic process that needs collaboration between government agencies, researchers, non-governmental organizations, and the public. Regular assessments and revisions based on evaluation outcomes assist to the continuing improvement of policies and boost their ability to solve critical environmental concerns.

ADAPTATION STRATEGIES

As the impacts of climate change become increasingly evident, the need for proactive adaptation strategies is paramount. These strategies aim to enhance resilience and minimize vulnerabilities in various sectors, ensuring communities, ecosystems, and economies can withstand and recover from the changing climate. Here are key adaptation strategies:

Climate-Resilient Infrastructure

1) ***Designing for Extreme Weather Events:*** Infrastructure projects should consider future climate projections, ensuring that buildings, roads, bridges, and other critical assets can withstand extreme weather events such as hurricanes, floods, and heatwaves.

2) ***Green Infrastructure:*** Incorporating green elements, such as green roofs, permeable pavements, and urban green spaces, helps manage stormwater, reduce urban heat islands, and enhance overall resilience.

Water Management and Conservation

1) ***Improved Water Storage and Distribution:*** Developing resilient water infrastructure, including efficient storage systems and distribution networks, helps communities manage water scarcity and adapt to changing precipitation patterns.

2) ***Water Conservation Programs:*** Implementing water conservation initiatives, including efficient irrigation practices and public awareness campaigns, helps ensure sustainable water use.

Ecosystem-Based Adaptation

1) ***Protecting and Restoring Ecosystems:*** Preserving and restoring natural ecosystems, such as forests, wetlands, and coastal areas, enhances biodiversity, supports natural adaptation processes, and provides crucial ecosystem services.

2) ***Biodiversity Conservation:*** Protecting diverse ecosystems contributes to ecosystem resilience, helping species adapt to changing conditions and maintaining the balance of ecological processes.

Crop Diversification and Sustainable Agriculture

1) ***Diversified Crop Varieties:*** Promoting the cultivation of a variety of crops that are resilient to different climate conditions enhances food security by mitigating the risks associated with climate-related disruptions.

2) ***Conservation Agriculture:*** Practices such as no-till farming and cover cropping improve soil health, water retention, and overall resilience in agriculture.

Early Warning Systems

1) ***Timely Alerts for Extreme Events:*** Developing and enhancing early warning systems for floods, hurricanes, heatwaves, and other extreme events helps communities prepare, evacuate, and respond effectively to minimize losses.

2) ***Community Outreach and Education:*** Ensuring that communities understand and respond to early warnings is critical for the success of these systems.

Climate-Resilient Coastal Management

1) ***Adaptive Coastal Infrastructure:*** Building and retrofitting coastal infrastructure to withstand sea-level rise, storm surges, and increased flooding protects coastal communities and reduces the risk of erosion.

2) ***Integrated Coastal Zone Management:*** Implementing comprehensive plans that balance economic activities, environmental conservation, and community needs in coastal areas ensures sustainable and resilient development.

Community-Based Adaptation

1) ***Local Knowledge Integration:*** Involving local communities in the planning and implementation of adaptation measures ensures that strategies are context-specific and draw on traditional knowledge.
2) ***Capacity Building:*** Strengthening the capacity of communities to cope with and adapt to climate impacts includes providing training, resources, and support for sustainable practices.

Health Preparedness and Response

1) ***Public Health Initiatives:*** Developing health strategies to address changing disease patterns, heat-related illnesses, and other climate-related health risks ensures communities are prepared for emerging challenges.
2) ***Healthcare Infrastructure Resilience:*** Strengthening healthcare infrastructure to withstand extreme weather events and ensuring access to medical services during emergencies is crucial for public health resilience.

Insurance and Risk Financing

1) ***Climate-Resilient Insurance Programs:*** Developing and promoting insurance programs that cover climate-related risks provides financial protection for individuals, businesses, and communities facing the impacts of extreme weather events.

2) ***Risk Financing Mechanisms:*** Establishing risk financing mechanisms, such as contingency funds and disaster risk reduction bonds, helps governments respond quickly and effectively to climate-related emergencies.

Energy Efficiency and Renewable Energy

1) *Diversification of Energy Sources:* Transitioning to renewable energy sources and improving energy efficiency reduces reliance on fossil fuels, mitigates climate impacts, and enhances energy security.

2) ***Resilient Energy Infrastructure:*** Designing energy infrastructure to withstand climate-related disruptions, such as extreme weather events and sea-level rise, ensures a reliable and resilient energy supply.

Education and Awareness Campaigns

1) ***Climate Literacy Programs:*** Implementing educational programs that enhance climate literacy at all levels of

society fosters a broader understanding of climate change and the importance of adaptation.

2) ***Community Workshops:*** Conducting workshops and awareness campaigns on climate change adaptation measures empowers communities to actively participate in resilience-building efforts.

Governance and Policy Integration

1) ***Mainstreaming Climate Adaptation:*** Integrating climate adaptation considerations into national and local policies across various sectors, including agriculture, water management, and urban planning, ensures a coordinated and effective response.

2) ***Multi-Stakeholder Collaboration:*** Facilitating collaboration between governments, businesses, civil society, and local communities ensures a collective and integrated approach to climate adaptation.

Technology Transfer and Innovation

1) ***Access to Climate-Resilient Technologies***: Facilitating the transfer of climate-resilient technologies to vulnerable regions ensures that communities have the tools needed to adapt to changing conditions.

2) ***Innovation Hubs:*** Supporting research and innovation in climate adaptation technologies fosters the development of new solutions and practices.

Regulation and Enforcement

1) ***Building Codes and Standards:*** Enforcing stringent building codes and standards that account for climate risks ensures that new constructions are resilient to extreme weather events.

2) ***Land Use Planning:*** Implementing regulations that guide sustainable land use planning helps prevent vulnerable development in high-risk areas.

Collaboration with Indigenous Knowledge

- ***Integration of Traditional Knowledge:*** Collaborating with indigenous communities and integrating traditional knowledge into adaptation strategies recognizes the value of local wisdom in understanding and responding to changing environmental conditions.

Successful climate change adaptation requires a combination of these strategies, tailored to the specific context of each region. A holistic and collaborative approach that involves governments, communities, businesses, and academia is essential for building a resilient and sustainable future in the face of climate uncertainty.

Building Resilient Infrastructure

Building resilient infrastructure is essential to ensure that communities can withstand and recover from the impacts of climate change, including extreme weather events, sea-level rise, and changing environmental conditions. Resilient infrastructure enhances the ability of societies to adapt and thrive in the face of evolving climate challenges. Here are key strategies for building resilient infrastructure:

Climate-Resilient Design

Future-Proofing: Designing infrastructure with climate projections in mind ensures that it can withstand changing weather patterns, increased temperatures, and extreme events over its lifespan.

Adaptive Capacity: Incorporating flexibility into design allows infrastructure to adapt to unforeseen climate impacts, reducing the need for extensive retrofitting in the future.

Green Infrastructure:

Natural Solutions: Integrating green infrastructure elements, such as green roofs, permeable pavements, and urban green spaces, helps manage stormwater, reduce the urban heat island effect, and enhance overall climate resilience.

Ecosystem-Based Approaches: Protecting and restoring natural ecosystems, such as wetlands and mangroves, can serve as natural buffers against storm surges and floods.

Risk Assessment and Planning

1) ***Vulnerability Assessments:*** Conducting comprehensive vulnerability assessments helps identify potential risks and vulnerabilities in existing and planned infrastructure projects.

2) ***Climate Risk Modeling:*** Utilizing advanced climate risk modeling tools aids in understanding the potential impacts of climate change on infrastructure assets.

Upgraded Transportation Networks

1) ***Flood-Resilient Roads and Bridges:*** Designing roads and bridges to withstand flooding and extreme weather events ensures the continued functionality of transportation networks during climate-related disruptions.

2) ***Alternative Transportation Modes:*** Encouraging and investing in alternative modes of transportation, such as public transit and non-motorized options, helps reduce reliance on vulnerable infrastructure.

Enhanced Water Management Systems

1) ***Smart Water Infrastructure:*** Implementing smart water management systems, including sensors and real-time monitoring, improves the efficiency of water distribution and helps mitigate water-related risks.

2) **Water Storage and Conservation:** Constructing resilient water storage systems and promoting water conservation practices enhance the capacity to manage water scarcity and floods.

Climate-Resilient Energy Infrastructure

1) *Diversification of Energy Sources:* Transitioning to a diversified and renewable energy mix reduces vulnerability to disruptions caused by extreme weather events and enhances energy security.

2) ***Resilient Power Grids***: Strengthening power grids and incorporating decentralized energy sources improves the resilience of energy infrastructure against climate-related disruptions.

Improved Building Codes

1) ***Climate-Adaptive Construction Standards***: Enforcing building codes that consider climate risks ensures that new

constructions are resilient to extreme weather events, such as hurricanes, floods, and heatwaves.

2) ***Elevated Structures:*** Elevating structures in flood-prone areas helps minimize damage and disruption during periods of increased flooding.

Community Engagement

1) ***Inclusive Decision-Making:*** Involving local communities in the planning and decision-making processes for infrastructure projects ensures that the infrastructure meets the specific needs and priorities of the people it serves.

2) ***Public Awareness Programs:*** Conducting public awareness programs about the importance of resilient infrastructure fosters community support and cooperation.

Investment in Resilience

1) ***Long-Term Funding Strategies:*** Developing sustainable funding mechanisms for infrastructure maintenance and upgrades ensures ongoing resilience against climate impacts.

2) ***Insurance and Risk Financing***: Utilizing insurance and risk financing mechanisms provides financial protection for infrastructure assets in the event of climate-related damages.

Cross-Sectoral Collaboration

1) ***Integrated Planning:*** Coordinating efforts across different sectors, such as transportation, water management, and energy, ensures a holistic and integrated approach to building resilient infrastructure.

2) ***Public-Private Partnerships:*** Encouraging collaboration between governments, private enterprises, and non-profit organizations enhances the collective capacity to build and maintain resilient infrastructure.

Climate-Responsive Urban Planning

Urban areas are particularly vulnerable to the impacts of climate change, but effective climate-responsive urban planning can mitigate risks, enhance resilience, and contribute to sustainable development. Here are key strategies for climate-responsive urban planning:

Compact and Sustainable Urban Design

- ***Mixed-Use Development:*** Encouraging mixed-use zoning reduces the need for extensive transportation, decreases the urban heat island effect, and promotes more sustainable and resilient urban spaces.

Compact Cities: Planning for compact, walkable cities minimizes the environmental footprint, reduces energy consumption, and enhances overall urban resilience.

Green Spaces and Urban Biodiversity

Urban Greening: Incorporating green spaces, parks, and urban forests within the urban fabric helps regulate temperatures, improve air quality, and enhance biodiversity.

- **Biodiversity Corridors:** Establishing corridors that connect green spaces allows for the movement of wildlife and supports ecosystem resilience within the urban environment.

Smart Infrastructure and Technology Integration

Smart Grids: Implementing smart energy grids, water management systems, and transportation networks enhances efficiency, reduces resource consumption, and improves the overall resilience of urban infrastructure.

- **Information and Communication Technologies (ICT):** Utilizing ICT for real-time data collection and analysis supports adaptive decision-making and improves emergency response capabilities.

Resilient Transportation Networks

Public Transit Systems: Investing in and expanding public transit systems reduces reliance on individual vehicles, mitigates traffic congestion, and enhances urban mobility during extreme events.

- ***Bicycle and Pedestrian Infrastructure***: Developing bike lanes and pedestrian-friendly pathways promotes sustainable and resilient urban transportation.

Climate-Responsive Building Codes

1) ***Green Building Standards:*** Enforcing green building codes that prioritize energy efficiency, sustainable materials, and climate-resilient designs contributes to the overall resilience of urban infrastructure.
2) ***Climate-Adaptive Construction:*** Incorporating climate-adaptive construction practices, such as elevated structures and flood-resistant designs, helps minimize the impact of extreme weather events.

Water Management and Flood Prevention

1) ***Green Stormwater Infrastructure:*** Implementing green stormwater infrastructure, such as rain gardens and permeable surfaces, helps manage stormwater runoff and reduces the risk of urban flooding.

2) ***Floodplain Protection:*** Designing urban areas with consideration for natural floodplains helps prevent urban flooding and protects critical infrastructure.

Community-Driven Planning

1) ***Participatory Urban Planning:*** Engaging communities in the urban planning process ensures that the needs and concerns of residents are considered, fostering a sense of ownership and cooperation.

2) ***Community-Based Resilience Programs***: Supporting community-led initiatives for climate resilience, such as neighborhood emergency response teams, enhances overall urban preparedness.

Adaptive Land Use Planning

1) ***Flexible Zoning Regulations:*** Establishing flexible zoning regulations allows for adaptive land use planning that considers changing climate conditions and potential shifts in population dynamics.

2) ***Resilient Coastal Management***: In coastal urban areas, integrating adaptive land use planning ensures sustainable development while minimizing risks associated with sea-level rise and storm surges.

Heat Mitigation Strategies

1) ***Cool Roof Initiatives:*** Promoting cool roof technologies helps reduce the urban heat island effect, lowering temperatures in densely populated areas.

2) ***Urban Greening for Shade:*** Planting trees strategically to provide shade in urban areas mitigates heat stress and contributes to a more comfortable urban environment.

Integrated Climate Action Plans

1) ***Climate-Responsive Policies:*** Integrating climate-responsive policies into overall urban development plans ensures that climate considerations are systematically addressed.

2) ***Carbon Neutral Goals***: Establishing goals for carbon neutrality and low-carbon urban development aligns urban planning with broader climate mitigation objectives.

Emergency Preparedness and Response

1) ***Urban Emergency Plans***: Developing and regularly updating emergency response plans for extreme weather events ensures a coordinated and effective response to climate-related disasters.

2) ***Evacuation Routes and Shelters:*** Identifying and communicating evacuation routes and establishing resilient

shelters enhances urban preparedness and protects vulnerable populations.

Climate Education and Awareness

1) ***Public Outreach Programs:*** Conducting educational programs and awareness campaigns about climate change impacts and adaptation measures fosters a climate-conscious urban population.

2) ***Capacity Building for Officials***: Providing training for urban planners and officials on climate-responsive urban planning practices ensures effective implementation and enforcement of climate-resilient policies.

Crisis Communication Systems

1) ***Effective Communication Channels:*** Establishing reliable communication systems that disseminate timely and accurate information during climate-related emergencies supports urban resilience and enhances public safety.

2) ***Community Engagement Platforms:*** Utilizing digital platforms and community networks for communication facilitates real-time updates and encourages community engagement in resilience efforts.

Transparency and Accountability

1) ***Open Data Initiatives:*** Implementing open data initiatives ensures that information about climate risks, urban planning decisions, and infrastructure projects is accessible to the public.
2) ***Accountability Mechanisms:*** Establishing mechanisms for accountability in urban planning processes ensures that decisions align with climate goals and community needs.

Collaboration with Stakeholders

1) ***Public-Private Partnerships:*** Collaborating with private enterprises, non-profit organizations, and research institutions strengthens the capacity for climate-responsive urban planning and promotes innovation.
2) ***Interdisciplinary Collaboration:*** Encouraging collaboration between urban planners, engineers, environmental scientists, and social scientists ensures a holistic and interdisciplinary approach to climate-responsive urban development.

Integrating these strategies into urban planning processes contributes to the creation of resilient, sustainable, and livable cities that can thrive in the face of climate challenges. The adoption of climate-responsive practices not only enhances the resilience of urban areas but also contributes to global efforts to mitigate the impacts of climate change.

FUTURE PROSPECTS

As the global community addresses the serious issues posed by climate change, several opportunities and chances emerge in the domains of mitigation and adaptation. The future trajectory is influenced by a combination of technical developments, legislative changes, international collaborations, and societal participation. Here are important future prospects in the current efforts to solve climate change:

Advancements in Renewable Energy Technologies

1) ***Breakthroughs in Solar and Wind:*** Ongoing research and development in solar and wind technologies are projected to deliver breakthroughs, making renewable energy sources more efficient, cost-effective, and scalable.

2) ***Energy Storage Solutions:*** Innovations in energy storage technologies, such as better batteries and grid-scale storage, will boost the reliability and stability of renewable energy systems.

Emergence of Climate-Smart Technologies

1) ***Climate-Resilient Agriculture:*** The development and use of climate-smart agricultural technology, including precision farming, drought-resistant crops, and sustainable irrigation systems, will contribute to food security in the face of changing climate circumstances.

2) ***Climate-Adaptive Infrastructure:*** Smart infrastructure equipped with sensors, artificial intelligence, and adaptive designs will become more ubiquitous, giving real-time data and adapting dynamically to climate-related concerns.

International Collaboration and Commitments

1) ***Strengthening Global accords:*** Enhanced coordination among nations to reinforce and exceed commitments contained in global accords like the Paris Agreement will be important for achieving common climate goals.

2) ***Technology Transfer and funding:*** Increased support for technology transfer and funding channels, particularly from developed to poor nations, will facilitate the adoption of sustainable solutions worldwide.

Policy Innovations and Legislation

1) ***Carbon Pricing Mechanisms:*** The growth and improvement of carbon pricing mechanisms, such as

carbon taxes and cap-and-trade systems, will motivate firms to cut emissions and invest in low-carbon technologies.

2) ***Ecosystem-Based Adaptation Policies:*** Governments are expected to implement policies that prioritize ecosystem-based adaptation, recognizing the value of natural solutions in strengthening resilience against climate impacts.

Decentralized Energy Systems

1) ***Community-Based Renewable Projects:*** The growth of community-based renewable energy projects, where local communities actively engage in and benefit from decentralized energy systems, will contribute to energy democratization and sustainability.

2) ***Microgrids and Off-Grid Solutions:*** Advances in microgrid technology and off-grid solutions will provide stable and resilient energy access to rural and underserved regions, reducing dependence on centralized energy networks.

Circular Economy and Sustainable Practices

1) ***Circular Business Models:*** The adoption of circular economy ideas, where items are intended for reuse,

recycling, and low waste formation, will gain significance across industries, minimizing environmental impact.

2) ***Regenerative Agriculture:*** The broad adoption of regenerative farming practices, focused at restoring soil health and sequestering carbon, will contribute to sustainable food supply and climate mitigation.

Technological Innovation for Carbon Removal

1) ***Direct Air Capture:*** Advancements in direct air capture technology will play a vital role in extracting carbon dioxide from the atmosphere, contributing to efforts to attain net-zero emissions.

2) ***Enhanced Weathering:*** Research and experimentation with enhanced weathering techniques, which entail accelerating natural processes that trap carbon in rocks, may offer scalable solutions for carbon removal.

Youth Activism and Public Engagement

1) ***young-Led Climate Movements:*** Continued young involvement and engagement in climate problems will influence political agendas, raise public awareness, and foster a feeling of urgency for climate action.

2) ***Climate Education Initiatives:*** The integration of climate education into school curricula and the development of

climate literacy will equip future generations to actively participate in sustainable practices and activism.

Climate-Resilient Finance and Investment

1) ***Green Finance Instruments***: The emergence of green finance instruments, such as green bonds and sustainable investment funds, will channel resources into initiatives with good environmental consequences.

2) ***Climate Risk Disclosure***: Increased emphasis on climate risk disclosure by businesses and financial institutions would promote transparency and accountability in analyzing climate-related financial risks.

Nature-Based Solutions for Adaptation

1) ***Blue Carbon Initiatives:*** The conservation and restoration of coastal ecosystems, such as mangroves and seagrasses, as blue carbon sinks will acquire recognized for their significance in climate adaption.

2) ***Urban Greening and Parks***: The growth of urban green spaces and the creation of climate-resilient parks will contribute to heat mitigation, biodiversity conservation, and better community well-being.

Digital Platforms for Climate Monitoring

1) ***Satellite and Sensor Technologies:*** Advancements in satellite and sensor technologies will increase the monitoring and understanding of climate change consequences, giving vital data for informed decision-making.

2) ***Blockchain for Climate Transparency:*** The use of blockchain technology to promote transparency in tracking and verifying carbon emissions reduction initiatives and sustainable supply chains will become increasingly popular.

Resilient Healthcare Systems

1) ***Climate-Adaptive Healthcare Infrastructure***: Future healthcare systems will integrate climate resilience concerns, preparing for the health implications of extreme weather events, shifting disease patterns, and other climate-related difficulties.

2) ***Telemedicine and Emergency Response:*** Technological developments, such as telemedicine and digital health platforms, will play a role in increasing healthcare accessibility and emergency response during climate-related disasters.

Adaptive Urban Planning and Design

1) ***Climate-Responsive Urban Design:*** Urban regions will increasingly adopt designs that incorporate climate hazards, incorporating green infrastructure, resilient building codes, and sustainable transportation options.

2) ***Smart Cities for Climate Resilience:*** The development of smart cities with integrated technology solutions will promote urban resilience, efficient resource utilization, and adaptive government.

Cross-Sectoral Collaboration

1) ***Multi-Stakeholder Partnerships***: Collaborative activities including governments, corporations, non-profit groups, academia, and local communities will become more widespread, supporting a holistic and inclusive approach to climate change.

2) ***Integration of Sustainable Development Goals (SDGs):*** Aligning climate action with larger sustainable development goals would create synergies and enable a comprehensive strategy to solving interrelated concerns.

Inclusive and Equitable Climate Solutions

1) ***Environmental Justice programs:*** Future climate programs will focus equitable solutions, addressing the

disproportionate impacts of climate change on vulnerable and marginalized people.

2) **Participatory Decision-Making:** Inclusion of various voices in decision-making processes will be vital for creating successful and just climate policy, ensuring that all viewpoints are considered.

While these prospects give promise for a more sustainable and resilient future, their realization depends on concentrated efforts, continuing innovation, and global cooperation. Governments, corporations, civil society, and individuals all play crucial roles in building a future where climate change is effectively managed, and societies are able to adapt to the challenges that lie ahead. The continued commitment to sustainable habits, informed policymaking, and the development of new solutions will be vital in attaining a climate-resilient and environmentally sustainable future.

Technological Innovations in Climate Science

Technological breakthroughs are transforming the field of climate science, giving researchers with new instruments to study, monitor, and understand the complexity of the Earth's climate system. These technologies contribute considerably to climate research, mitigation initiatives, and adaption techniques. Here are major technological developments in climate science:

Earth Observation Satellites

1) ***High-Resolution Imaging:*** Advanced satellite systems equipped with high-resolution imaging capabilities allow scientists to monitor changes in land cover, deforestation, glacier melt, and other vital indicators, giving valuable data for climate studies.

2) ***Remote Sensing Technologies:*** Utilizing numerous remote sensing technologies, including LiDAR and radar, enables the assessment of air factors, sea surface temperatures, and changes in vegetation with unparalleled accuracy.

Climate Modeling Supercomputers

1) ***High-Performance Computing:*** Supercomputers equipped with powerful modeling software mimic complicated climate processes, enabling scientists anticipate future climate scenarios and comprehend the implications of human actions on the environment.

2) ***Ensemble Modeling:*** Ensemble modeling, which incorporates numerous climate models, boosts the trustworthiness of climate projections and provides a more comprehensive view of likely future outcomes.

Machine Learning and Artificial Intelligence (AI)

1) ***Pattern Recognition:*** Machine learning algorithms examine enormous datasets to discover patterns and trends in climate data, aiding in the detection of abnormalities, severe events, and potential climate change indicators.

2) ***Climate Prediction Models:*** AI is increasingly integrated into climate prediction models, boosting their accuracy and enabling real-time forecasting of weather events and long-term climate trends.

Climate Data Analytics Platforms:

1) ***Big Data Analytics:*** Platforms that process and analyze vast climate datasets enable researchers to extract insights, identify patterns, and assess the implications of climate change on regional and global scales.

2) ***Visualization technologies:*** Advanced data visualization technologies assist explain complicated climate information to policymakers, stakeholders, and the public, enabling greater understanding and informed decision-making.

Ocean Monitoring Technologies

1) ***Ocean Gliders and Buoys***: Autonomous underwater gliders and buoys equipped with sensors capture real-time data on

ocean temperature, salinity, and currents, adding to the knowledge of ocean dynamics and climate patterns.

2) ***Satellite Altimetry:*** Satellite altimetry measures variations in sea surface height, providing essential information on ocean circulation, sea level rise, and the distribution of heat in the seas.

Ice Monitoring and Remote Sensing

1) ***Ice-Tracking Satellites:*** Specialized satellites equipped with radar and other sensors monitor changes in polar ice caps, glaciers, and sea ice, helping scientists analyze the rate of ice melt and its impact on sea level rise.

2) ***Ice Core Analysis:*** Technological improvements in ice core analysis techniques provide insights into historical climate conditions, allowing researchers to reconstruct previous climate fluctuations and analyze long-term trends.

Carbon Monitoring Technologies

1) ***Satellite-based Carbon Observatories:*** Satellites equipped with spectrometers and sensors monitor atmospheric carbon dioxide concentrations, aiding in the identification of sources and sinks of carbon, as well as tracking global carbon cycles.

2) ***Ground-based Monitoring Networks:*** Advanced ground-based networks of sensors and observatories continually measure greenhouse gas emissions, giving real-time data for climate models and policy assessments.

Climate Resilience Decision Support Systems

1) ***Integrated climatic Risk Assessments:*** Decision support systems integrate climatic data, socioeconomic aspects, and vulnerability assessments to help communities, corporations, and governments plan and implement climate-resilient initiatives.

2) ***Scenario Planning Tools:*** Advanced scenario planning tools use climate models to estimate probable future consequences, supporting policymakers and planners in building adaptive solutions for various climate scenarios.

Next-Generation Weather Forecasting

1) ***Numerical Weather Prediction Models***: Continual developments in numerical weather prediction models, driven by improved computing power and data assimilation techniques, lead to more accurate short-term weather forecasts and severe weather predictions.

2) ***High-Resolution Ensemble Forecasting:*** Ensemble forecasting, integrating numerous model runs with modest

differences, boosts the reliability of weather predictions and improves the understanding of uncertainty in forecasting.

Climate Monitoring from Space Stations

1) ***Experiments on International Space Station (ISS):*** Instruments deployed on the ISS collect significant data on Earth's environment, examining phenomena such as auroras, cloud formation, and atmospheric composition from a unique vantage point.

2) ***International Collaboration in Space Research***: Collaborative activities involving space agencies and international partners lead to a more comprehensive understanding of global climate patterns and trends.

Predictions for Climate Trends and Challenges

Anticipating future climate trends and difficulties includes examining a number of elements, including scientific breakthroughs, legislative decisions, human habits, and the Earth's intrinsic response to continuing climate change. While particular estimates may differ, certain broad themes and problems are anticipated to affect the future climate landscape:

Continued Global Temperature Rise

- ***Warming Trend***: The general global temperature is predicted to continue rising due to the accumulation of greenhouse gases in the atmosphere, leading to more frequent and intense heatwaves, with possible repercussions on ecosystems, agriculture, and human health.

Changes in Precipitation Patterns

- ***Shifts in Rainfall and Snowfall***: Climate models imply adjustments in precipitation patterns, including changes in the frequency and intensity of rainfall events, shifts in seasonal precipitation, and variations in snowfall patterns, influencing water supplies and ecosystems.

Sea Level Rise and Coastal Challenges

1) ***Ongoing Sea Level Rise:*** Melting glaciers and ice caps, together with the thermal expansion of seawater, contribute to rising sea levels, creating dangers to coastal residents, infrastructure, and ecosystems.

2) ***Increased Coastal Flooding:*** Sea level rise, coupled with storm surges, will enhance the risk of coastal flooding

occurrences, hurting low-lying areas and mandating adaptive measures to protect vulnerable coastal regions.

Intensification of Extreme Weather Events

1) ***More Intense Hurricanes and Cyclones:*** Warmer ocean temperatures are projected to fuel more intense tropical storms and hurricanes, leading to greater intensity and possible damage in coastal communities.

2) ***Changes in Extreme Precipitation:*** Extreme rainfall events, leading to floods and flash floods, are anticipated to become more frequent in many places, damaging populations and infrastructure.

Biodiversity Loss and Ecosystem Disruptions

1) ***Species Extinction and Migration:*** Climate-induced changes in temperature and ecosystems may lead to shifts in species distributions, with some risking extinction, while others may migrate to new places, altering biodiversity and ecological balance.

2) ***Disruption of Ecosystem Services***: Alterations in climate conditions can disrupt vital ecosystem services, such as pollination, water purification, and regulation of disease vectors, harming human well-being and biodiversity.

Ocean Acidification and Coral Reef Decline

1) ***Impact on Marine Ecosystems:*** The absorption of excess carbon dioxide by the oceans leads to acidification, affecting marine life, particularly creatures with calcium carbonate shells, and contributing to the decline of coral reefs.

2) ***Loss of Biodiversity in Oceans:*** Coral bleaching and deterioration of coral reef ecosystems, worsened by rising temperatures, represent enormous challenges to marine biodiversity and the lives of communities depending on marine resources.

Food Security Challenges

1) ***Shifts in Agricultural Productivity***: Changes in temperature and precipitation patterns may effect crop yields and agricultural productivity, leading to issues in food security, particularly in places that rely primarily on rain-fed agriculture.

2) **Water Scarcity and Agricultural Practices**: Increasing water scarcity, coupled with changing precipitation patterns, demands creative water management measures to support agriculture and assure food security.

Increased Frequency of Wildfires

1) ***Wildfire Intensity and Frequency:*** Rising temperatures, extended droughts, and changes in vegetation patterns contribute to an increased risk of wildfires in diverse locations, causing dangers to ecosystems, communities, and air quality.

2) ***Challenges for Fire control:*** The escalation of wildfires needs increased fire control measures, including early detection, community preparedness, and sustainable land management practices.

Migration and Climate-Induced Displacement

1) ***Environmental Refugees:*** Climate change impacts, such as sea-level rise, extreme weather events, and resource scarcity, may contribute to increased migration and displacement, leading to social, economic, and political issues.

2) ***International Cooperation on Climate Migration:*** Addressing climate-induced displacement requires international cooperation to design policies that protect the rights and well-being of displaced communities.

Infrastructure Vulnerabilities and Resilience Needs

1) ***Adaptation of Critical Infrastructure:*** Rising temperatures, extreme weather events, and sea-level rise pose challenges to the resilience of infrastructure, demanding investments in adaptive measures and the incorporation of climate resilience into urban design.

2) ***Energy Infrastructure shift:*** The shift to renewable energy sources and the development of resilient energy infrastructure are vital for reducing the impacts of climate change on power networks and guaranteeing reliable electricity availability.

Water Scarcity and Stress

1) ***Increasing Water Stress:*** Changes in precipitation patterns, melting glaciers, and population increase lead to water scarcity in diverse places, requiring sustainable water management techniques, conservation initiatives, and the development of alternative water sources.

2) ***Conflicts Over Water Resources:*** Competition for decreasing water resources may intensify geopolitical tensions and lead to conflicts, underlining the significance of equitable and cooperative water management measures.

Global Policy Responses and Climate Agreements

1) ***Strengthening Climate Policies:*** The effectiveness of global efforts to prevent climate change depends on the

adoption of ambitious and enforced climate policies, including carbon price mechanisms, emissions reduction targets, and adaptation plans.

2) ***Enhanced International Collaboration:*** Continued collaboration among states, companies, and civil society is vital for fulfilling the goals specified in international agreements such as the Paris Agreement and increasing collective efforts to combat climate change.

Technological Solutions and Innovation

1) ***Acceleration of Clean technology:*** Rapid breakthroughs in clean energy technology, carbon capture and storage, and sustainable practices are vital for achieving carbon neutrality and minimizing the impacts of climate change.

2) ***Technological Innovation for Adaptation:*** Innovation in climate-resilient technologies, including infrastructural solutions, agricultural techniques, and disaster preparedness systems, will be crucial for boosting adaptive capacity.

Social and Behavioral Change

1) ***Climate Education and Awareness:*** Increasing climate literacy, encouraging awareness, and promoting sustainable habits are vital to creating resilience at the community level and generating public support for climate legislation.

2) ***Community-Led Climate Initiatives:*** Grassroots movements and community-led initiatives play a crucial role in pushing climate action, advocating for sustainable practices, and promoting resilience within local communities.

Global Economic Impacts and Opportunities

1) ***Transition to a Green Economy:*** The shift towards a green economy, typified by sustainable practices, circular business models, and investments in renewable energy, gives prospects for economic growth, job creation, and innovation.

2) ***Costs of inactivity:*** The economic costs of inactivity on climate change, including damage to infrastructure, health impacts, and disruptions to supply networks, underline the need of proactive climate mitigation and adaptation measures.

While these forecasts offer insights into potential future trends and challenges, the exact consequences will depend on the collaborative actions made by governments, organizations, communities, and individuals. The course of climate change and its repercussions can be altered by mitigation efforts, adaptation methods, and the ability of society to shift towards sustainable and resilient practices. Proactive and coordinated efforts are required to confront the multiple challenges posed by climate change and establish a more sustainable future for generations to come.

CONCLUSION: NAVIGATING THE COMPLEX TERRAIN OF CLIMATE CHANGE

The journey through the intricate landscape of climate change reveals a multifaceted tapestry woven by the interplay of scientific insights, technological innovations, policy endeavors, and societal dynamics. As we stand at the intersection of unprecedented challenges and transformative opportunities, the need for collective action and resolute commitment becomes increasingly evident.

The urgency for immediate action emanates from the stark realities painted by the canvas of climate science. The Earth, our shared home, is undergoing profound changes, manifested in rising temperatures, altered precipitation patterns, and the intensification of extreme weather events. The consequences ripple through ecosystems, impacting biodiversity, threatening food security, and testing the resilience of communities worldwide.

Our exploration of technological innovations in climate science illuminates a path forward. Advanced satellite systems, machine learning algorithms, and climate modeling supercomputers empower scientists with unprecedented capabilities to understand, monitor, and predict the intricacies of the climate system. These technological tools not only deepen our comprehension of climate phenomena but also offer crucial insights for crafting effective mitigation and adaptation strategies.

Predictions for future climate trends and challenges present a call to action. The trajectory ahead is marked by the continuation of global temperature rise, shifts in precipitation patterns, and the escalating impacts on ecosystems, agriculture, and human societies. Sea level rise, intensified extreme weather events, and the specter of climate-induced displacement demand comprehensive solutions rooted in resilience, innovation, and global cooperation.

In the quest for sustainability, global policies and agreements emerge as beacons of hope. The Paris Agreement, a testament to international collaboration, sets the stage for ambitious climate action and emissions reduction commitments. Strengthened climate policies, coupled with technological advancements, offer a roadmap for a transition to a low-carbon future.

However, the narrative of climate change extends beyond the realm of policies and technologies; it is deeply intertwined with the choices we make, the values we uphold, and the collective responsibility we bear. The conclusion drawn from our exploration is not a finale but a call to engagement. It is a recognition that the power to shape the future lies in the hands of individuals, communities, businesses, and governments alike.

As we navigate this complex terrain, the imperative for sustainable practices, equitable solutions, and inclusive decision-making comes to the fore. Education and awareness emerge as catalysts for change, empowering communities to adopt climate-resilient behaviors and advocate for a sustainable future. Grassroots movements, fueled by the passion of individuals and the collaboration of diverse stakeholders, signify the strength of collective action in the face of global challenges.

In the pursuit of a resilient and sustainable future, the conclusion drawn is one of shared responsibility and unwavering commitment. It is an acknowledgment that the choices we make today echo across generations, influencing the legacy we leave for those who will inherit the Earth. The story of climate change is not predetermined; it is a narrative being written by our actions, innovations, and collaborations.

In concluding this exploration, we recognize that the journey is ongoing, and the challenges ahead demand perseverance, creativity, and a unity of purpose. The path to a sustainable future requires a continuous dialogue, a commitment to environmental stewardship, and a shared vision of a world where the balance between humanity and the planet is restored.

The concluding chapter is an invitation – an invitation to embark on a collective journey towards a resilient, sustainable, and harmonious coexistence with the Earth we call home.

Summarizing Key Findings and Recommendations for Action

Key Findings

Climate Science Insights

The Earth is experiencing a warming trend due to human-induced greenhouse gas emissions.

Advanced climate science technologies, including satellites, supercomputers, and AI, enhance our understanding of climate patterns and facilitate more accurate predictions.

Technological Innovations

Earth observation satellites and machine learning contribute to real-time monitoring and analysis of climate-related changes.

Climate modeling supercomputers enable simulations for predicting future climate scenarios.

Advanced technologies in ice monitoring, carbon tracking, and ocean observations provide critical data for climate research.

Predictions for Future Climate Trends

Global temperature rise and altered precipitation patterns are expected to continue.

Sea level rise, intensified extreme weather events, and biodiversity loss pose significant challenges.

The world faces increased risks of wildfires, water scarcity, and climate-induced displacement.

Policy Landscape

Global climate agreements, notably the Paris Agreement, highlight international commitment to emissions reduction and climate resilience.

Strengthened climate policies and innovative solutions are essential for achieving sustainability goals.

Societal Dynamics

Public awareness and grassroots movements play a crucial role in advocating for sustainable practices.

Climate education and community engagement are pivotal for fostering a sense of shared responsibility.

Recommendations for Action

Accelerate Transition to Renewable Energy

Invest in and incentivize the development of renewable energy technologies.

Implement policies supporting the transition to a low-carbon energy sector.

Enhance Climate Resilience

Incorporate climate-resilient design principles in infrastructure projects.

Develop and implement climate adaptation plans at local and national levels.

Advance Climate Education

Integrate climate education into school curricula.

Promote public awareness campaigns to enhance climate literacy.

Support Sustainable Practices

Encourage sustainable agriculture and land-use practices.

Foster circular economy principles to minimize waste and resource consumption.

Strengthen International Collaboration

Collaborate on technological innovations for climate monitoring and mitigation.

Support global initiatives for climate resilience and emissions reduction.

Promote Inclusive Decision-Making

Involve local communities in climate planning and decision-making.

Prioritize inclusive policies that address the needs of vulnerable populations.

Invest in Climate-Resilient Infrastructure

Allocate resources for the development of climate-resilient infrastructure.

Implement building codes that consider climate risks and extreme events.

Drive Innovation for Climate Solutions

Invest in research and development of climate-resilient technologies.

Support startups and initiatives focused on sustainable practices and clean technologies.

Foster Sustainable Urban Planning

Implement climate-responsive urban planning strategies.

Develop green spaces and promote sustainable transportation options in urban areas.

Advocate for Ambitious Climate Policies

Encourage governments to set and enforce ambitious emission reduction targets.

Advocate for the implementation of carbon pricing mechanisms.

Promote Nature-Based Solutions

Support conservation and restoration of ecosystems for carbon sequestration.

Implement blue carbon initiatives to protect coastal ecosystems.

Build Climate-Resilient Communities

Develop community-led resilience programs and emergency response plans.

Invest in social infrastructure to support vulnerable communities.

Embrace Sustainable Business Practices

Encourage businesses to adopt sustainable and eco-friendly practices.

Support corporate initiatives for carbon neutrality and environmental responsibility.

Facilitate Climate-Responsive Healthcare

Incorporate climate considerations into healthcare infrastructure planning.

Develop adaptive healthcare systems to address climate-related health challenges.

Encourage Green Finance and Investments

Support green finance instruments and sustainable investment practices.

Encourage financial institutions to disclose climate-related risks in their portfolios.

In conclusion, addressing the challenges posed by climate change requires a comprehensive and collaborative approach. These key findings and recommendations underscore the importance of individual and collective actions, technological innovation, and policy advocacy to create a sustainable and resilient future for generations to come. By embracing these recommendations and working together on a global scale, we can navigate the climate challenge and build a world that balances the needs of humanity with the health of our planet.

ACKNOWNLEDGEMENT

In this book, we are taken on a journey through the complex web of climate change, one that is full of obstacles, opportunities for growth, and the shared goal of a more sustainable future. This story has been a group effort, and we are all grateful to everyone who has helped bring it to fruition.

We are incredibly grateful to the scientific community for their unwavering dedication to understanding the intricacies of climate science. Your priceless insights have served as the compass points for our investigation, researchers, climatologists, and environmental experts.

Particular thanks are due to the technical trailblazers whose work has changed the face of climate research. We owe a debt of gratitude to the algorithms that analyze massive datasets and the satellites that orbit the Earth for helping us understand the complexities of our changing climate.

The lawmakers and campaigners advocating climate action deserve praise for their commitment to establishing a policy landscape that encourages sustainability. Your tireless efforts lay the way for genuine change and global cooperation.

Your unwavering commitment and enthusiasm motivate people all throughout the globe who are taking part in climate activism and grassroots initiatives. Your voices resound through these pages, reminding us that the battle against climate change is a collective enterprise.

The educators and communicators propagating climate literacy are acknowledged for their essential role in creating awareness and understanding. As a result of your work, understanding becomes a driving force behind significant progress.

Our deepest gratitude goes out to all who have made a commitment to innovation, resilience, and sustainable practices. Your activities, whether great or small, contribute to the tapestry of solutions needed to handle the challenges ahead.

To those who campaign for diversity, equity, and justice in the face of climate effects, your commitment to ensuring that no one is left behind is a beacon guiding our path toward a more just and sustainable society.

We offer gratitude to the worldwide community, understanding that the collaborative spirit of governments, corporations, and individuals is crucial in the collective response to climate change. It is through shared responsibility and cooperation that we can construct a resilient and sustainable future.

Lastly, to the readers who embark on this trip with us, your participation and commitment to comprehending the nuances of climate change are the driving force behind these words. May this exploration inspire meaningful action and light the flame of change in the hearts of all who turn these pages.

In appreciating each contributor to our shared narrative, we realize that the road toward a sustainable future is one we undertake together. As we move forward, may our combined efforts continue to construct a world where harmony between humans and the planet is restored.